Y0-BJK-949

LIFE IN THE MILITARY

LIFE IN THE
US MARINE CORPS

by Susan E. Hamen

BrightPoint Press

San Diego, CA

BrightPoint Press

© 2021 BrightPoint Press
an imprint of ReferencePoint Press, Inc.
Printed in the United States

For more information, contact:
BrightPoint Press
PO Box 27779
San Diego, CA 92198
www.BrightPointPress.com

ALL RIGHTS RESERVED.

No part of this work covered by the copyright hereon may be reproduced or used in any form or by any means—graphic, electronic, or mechanical, including photocopying, recording, taping, web distribution, or information storage retrieval systems—without the written permission of the publisher.

LIBRARY OF CONGRESS CATALOGING-IN-PUBLICATION DATA

Names: Hamen, Susan E., 1976- author.
Title: Life in the US Marine Corps / Susan E. Hamen.
Description: San Diego : ReferencePoint Press, [2021] | Series: Life in the military | Includes bibliographical references and index. | Audience: Grades 10-12
Identifiers: LCCN 2020002428 (print) | LCCN 2020002429 (eBook) | ISBN 9781682829752 (hardcover) | ISBN 9781682829769 (eBook)
Subjects: LCSH: United States. Marine Corps. | United States. Marine Corps--Vocational guidance.
Classification: LCC VE23 .H2915 2020 (print) | LCC VE23 (eBook) | DDC 359.9/6120973--dc23
LC record available at https://lccn.loc.gov/2020002428
LC eBook record available at https://lccn.loc.gov/2020002429

CONTENTS

AT A GLANCE	4
INTRODUCTION PROVIDING RELIEF	6
CHAPTER ONE HOW DOES A PERSON BECOME A MARINE?	12
CHAPTER TWO WHAT JOBS ARE AVAILABLE?	32
CHAPTER THREE WHAT IS THE LIFE OF A MARINE LIKE?	46
CHAPTER FOUR WHERE DO MARINES SERVE?	58
Glossary	74
Source Notes	75
For Further Research	76
Index	78
Image Credits	79
About the Author	80

AT A GLANCE

- The US Marine Corps (USMC) is a branch of the US military. It is part of the US Navy. People who serve in the USMC are called marines.

- Marines defend the country on land, in the air, and at sea. Nearly 300,000 people serve in the USMC.

- Most marines serve full-time. Others are part of the Marine Corps Reserve. They serve part-time.

- Marine recruits must pass a medical test and an entrance exam. They take a physical fitness test too.

- Recruits go through a training period. It lasts thirteen weeks. It is called boot camp.

- There are two boot camp locations. One is in Parris Island, South Carolina. Male and female recruits go there. The other is in San Diego, California. Only male recruits go there.

- The USMC offers more than 180 different jobs.

- Some marines serve in the United States. Others serve in other countries. They fight in combat. They also help people in times of crisis.

INTRODUCTION

PROVIDING RELIEF

On September 20, 2017, a hurricane hit Puerto Rico. It was called Hurricane Maria. It devastated the island. It brought strong winds. The winds reached 155 miles per hour (250 km/h). They took down power lines and damaged buildings.

After the storm, many people could not find food or clean water. Fallen trees and

A US Marine helps clear roads in Puerto Rico after a deadly hurricane hit the island in 2017.

buildings blocked the roads. People did not have fuel or power. Cell phones did not work. Luckily, help arrived.

A US Navy helicopter drops off supplies on the USS Kearsarge. *The US Marines and the US Navy worked together on relief efforts after Hurricane Maria.*

A US Navy ship arrived in Puerto Rico five days after the storm. The ship was

8

called the USS *Kearsarge*. Marines were on board the ship. Their job was to help with relief efforts. They distributed food, water, and medical supplies. They also helped clean up streets and repair power lines. They cleared airfields. This allowed airplanes to land safely. The planes carried much-needed supplies.

The marines worked around the clock to offer aid. They helped restore power to hospitals in two cities. Marine helicopter pilots flew doctors to areas that were hard to access. The doctors treated people who needed medical help.

WHO ARE THE MARINES?

Marines serve in the US Marine Corps (USMC). The USMC is part of the US Navy. It is one of the smallest branches of the US military.

The focus of the Marine Corps is combat. But marines also help people in times of crisis. Both men and women can serve in the Marines. Sometimes the Marine Corps works together with the navy.

Marines have specialized jobs. They mainly work on land and at sea. But some are involved in air combat. Marines serve in many places, including deserts and jungles.

Marines use camouflage paint to blend into their surroundings.

They may have to work in extreme weather.

They go wherever they are needed.

CHAPTER ONE

HOW DOES A PERSON BECOME A MARINE?

Nearly 300,000 people serve in the Marine Corps. Some are active duty. They serve full-time. Others serve in the Marine Corps Reserve. They work part-time. They may have a job outside of the Marine Corps. They may be called to serve full-time during an emergency.

Recruits swear an oath to become a part of the Marines. They promise to defend the United States.

The process of joining the Marines is called enlisting. Some people join the Marines as enlisted troops. Others join as officers. Officers usually fill leadership roles. They are ranked higher than enlisted troops.

Enlisted troops can take a course and train to become low-ranking officers, called warrant officers.

They can train and lead enlisted troops. They may organize missions.

People must meet requirements to join the Marines. They must have a high school diploma. They need to be a legal US resident. This means they can legally live and work in the United States. Applicants must also be between the ages of seventeen and twenty-eight. Seventeen-year-olds may join with their parents' permission. Applicants also need to meet height and weight standards.

OFFICER REQUIREMENTS

People who want to enlist as an officer must meet additional requirements. They need to have a four-year college degree. They must be between the ages of eighteen and twenty-seven.

There are a few ways to become a Marine Corps officer. People can start the process while in college. They take part in a program. It is called the National Reserve Officer Training Corps. The students become officers when they graduate from college.

Another option is Officer Candidate School (OCS). OCS is at Marine Corps Base Quantico in Virginia. People can go through OCS after college. Or they can go in the summers while still in college.

The US Naval Academy (USNA) also trains people to become officers. This school is in Annapolis, Maryland. Getting into the USNA is difficult.

A Marine Corps sergeant shows Officer Candidate School students how to use a compass and navigate in the wilderness.

Applicants need a recommendation. The recommendation must come from a US representative or senator. The US president and vice president can also recommend applicants. Those who are accepted go to the school for four years. The US Navy pays the students' education costs. It also covers living expenses. USNA students can choose to go into the Marine Corps. They must go through four weeks of special training. They learn leadership skills. They must pass a combat test and a fitness test.

ENLISTING

Recruiters guide people through the enlistment process. Recruiters are marines. They explain what it is like to be in the Marine Corps. They answer questions and offer advice.

People who enlist are called **recruits**. Their first step is to sign an enlistment contract. This contract explains how long

The US Naval Academy was originally called the Naval School. It was founded in 1845.

someone will serve. Recruits must serve at least four years. The first time a person enlists, there is a chance her service will be extended. She may be called to serve up to four more years.

Next, recruits must pass a medical test. Doctors examine them to make sure they are healthy and fit for the military.

People can take the ASVAB during their junior or senior year of high school.

Recruits also take an entrance exam. It is called the Armed Services Vocational Aptitude Battery (ASVAB) test. The ASVAB is required for every branch of the military. It measures people's skills in ten subject areas. It assesses math, science, and language skills. It also measures mechanical and electrical skills. People have separate scores in each area. These scores help them find jobs. Some jobs within the USMC require higher ASVAB scores. Recruits who do well on all parts of the ASVAB have many job choices.

BOOT CAMP

The next step is basic training. This training is also called boot camp. Recruits in other branches of the military go through this training too. But Marine boot camp is tougher than any other military boot camp. Lieutenant Colonel Ty Kopke is a drill instructor. He shouts orders to recruits. Recruits must obey these orders. Kopke says, "[Boot camp is] supposed to prepare you for the challenges that lie beyond."[1]

Boot camp happens at two locations. These locations are called Marine Corps Recruit Depots. One depot is on Parris

PARRIS ISLAND

Parris Island became a US naval station in 1889. The island is off the coast of South Carolina. It is about 4 miles (6 km) long. It became a Marine Corps Recruit Depot in 1915. The first mixed-gender class graduated in 2019. The class included men and women. This was the first time male and female Marine recruits trained together. The USMC usually trains them separately.

Island in South Carolina. Another is in San Diego, California. Female recruits go to Parris Island. Recruits can write letters to their families during boot camp. But they cannot contact their families any other way.

THE THREE PHASES

The first part of boot camp is called Phase One. This phase lasts four weeks. Recruits get their first haircut. They also get their gear and uniforms. They go through a medical and dental exam.

Recruits learn about the Marine Corps' values. These values include honor, courage, and commitment. Recruits study

Marine Corps history. They also learn how to handle weapons. They practice aiming and firing rifles. They work on hand-to-hand combat skills. They learn first aid skills too.

Swim training also happens in Phase One. Recruits learn water survival skills. They might need to swim on a mission. They need to be prepared for this. Recruits swim in full uniform. They practice treading water. They learn how to shed gear and use flotation devices.

Recruits take a test in Phase One. It is called the Initial Strength Test (IST). The IST includes pull-ups and crunches. Recruits run 1.5 miles (2.4 km). They must meet certain minimum standards. The standards depend on the job they want to pursue. Some jobs require heavy lifting. People who want to do these jobs must do at least three pull-ups. They have to do forty-four

THE FIRST HAIRCUT

Male recruits get a haircut on the first day of boot camp. Their hair is buzzed short. Sergeant Christopher Bess is a drill instructor. He explains why all male recruits get a buzz cut: "When . . . they all look the same they'll start to act the same. It also helps them grow as a team." Female recruits do not have to get a buzz cut. But they must keep their hair neatly groomed.

Quoted in Corporal Liz Gleason, "Weekly Haircuts Create Uniformity," US Marine Corps, March 7, 2013. www.mcrdsd.marines.mil.

crunches in two minutes. All recruits must complete the run in thirteen minutes and thirty seconds. The recruits have to lift **ammunition** cans too. These are boxes that hold ammunition, or ammo. They weigh 30 pounds (14 kg). Recruits must do forty-five ammo can lifts.

Some recruits will not have jobs that require much heavy lifting. Some of their IST requirements are different. Lifting is not part of the test. Men must do two pull-ups. Women have to do a flexed, twelve-second arm hang. The crunches requirements are the same. The run standards are also the

Marine recruits must swim 27 yards (25 m) with their packs.

same for men. Women have fifteen minutes to complete the 1.5-mile (2.4-km) run.

Recruits who pass Phase One move on to Phase Two. This phase takes five weeks. Recruits become skilled shooters. They learn navigation and survival skills.

Marines practice putting on gas masks before gas chamber training.

One of the biggest challenges in Phase Two is gas chamber training. Recruits enter a closed room. Tear gas is released into the room. This gas irritates the recruits' eyes. It also hurts their throat and lungs. They may have difficulty breathing. Recruits learn how to use a gas mask. Gas masks help them breathe.

Phase Three is the last part of basic training. It lasts four weeks. Recruits learn how to drive military trucks. They are tested on Marine Corps history and other knowledge. The most difficult part of this phase is the Crucible. This event lasts fifty-four hours. Recruits go through a combat exercise. The sounds of gunfire are played from speakers. Recruits work in teams. They go through obstacles. They get little food. They only get eight hours of sleep. They march about 40 miles (64 km).

Staff Sergeant Jasmine Rodgers is a drill instructor. She says, "The Crucible is actually seeing everything that has been taught to them come to life. It shows the best and the worst from them, and by the end of it, they are better for it."[2]

Working in the Marines is physically challenging. Boot camp prepares recruits for these challenges. One recruit explained, "You go through a really rough time. You start thinking it's hot, I'm thirsty, my arms haven't felt this bad in my whole life."[3] But getting through boot camp can be rewarding. Recruits who pass all three

Marines have to complete many difficult obstacles during the Crucible. The obstacles test their strength and endurance.

phases graduate. They go through a special ceremony. Their families may come to watch the ceremony. This is a joyful time. The recruits' hard work has paid off. They have become marines.

CHAPTER TWO

WHAT JOBS ARE AVAILABLE?

After boot camp, marines train to work a specific job. The Marine Corps offers more than 180 different jobs. Many marines are stationed at bases around the world. They are ready to respond when needed. Other marines live and work aboard ships. They help other countries in times of crisis. They might help

Each marine has a specific job. Some marines are trained to load and secure weapons such as rockets into aircraft.

countries recover after a hurricane or other natural disaster.

The Marine Corps uses navy ships. But it has its own aircraft, weapons, and supplies. Many marines maintain or work with these items. Some marines work as **infantry**.

Others are aircraft mechanics or air traffic controllers. These are just some of the many roles marines have.

INFANTRY

All marines are called riflemen. Each marine receives special training on how to

LINGUISTS

Some marines work in other countries. People in another country may speak a different language. Marines need to be able to communicate with them. Some marines are linguists. They can speak another language. They translate foreign languages into English. The marines have linguists for more than fifty languages. Linguists take language classes. They must pass tests that prove they are fluent in the language.

use a rifle. If needed, any marine can be called upon to serve in the infantry. The infantry involves ground combat. These marines fight on foot. Their goal is to seize and control land. They try to defeat enemies. These marines specialize in hand-to-hand combat.

   Many types of infantry jobs are available. Some marines are scout snipers. A scout sniper is a highly specialized position. Marines practice firing rifles on shooting ranges. They fire at targets from hidden positions. They shoot from far away. That way, the enemy might not be able to find them. Scout snipers also spy on enemies. They gather information about enemy forces. The military uses this information to plan attacks.

Peter James Kiernan is a former Marine scout sniper. He went through sniper school. He says, "Scout snipers' missions require them to be . . . miles away from friendly support. . . . The [sniper] course teaches what it's like to be alone and unafraid."[4]

Snipers must have excellent aim. They must have good vision. They can hit a target without hurting nearby people. Snipers have

Marine scout snipers spend much of their time training on shooting ranges.

to be **stealthy**. They must be able to easily hide themselves. They learn how to blend in with their surroundings.

TANKS

The Marine infantry also uses tanks. Some marines are part of the tank crew. They are trained to use these vehicles. Four crew

Marines on a tank crew must be able to work well together.

members are assigned to each tank. One person drives the tank. The driver also does any needed maintenance. Another crew member is the tank gunner. The gunner helps prepare equipment. This person also fires the tank's weapons. A third crew member is the loader. This person loads ammunition into the tank's weapons. A fourth crew member is the tank commander. This person leads the crew.

AIRCRAFT

The Marine Corps has its own air force. It is one of the largest air forces in the world. The Marine Corps maintains and flies different types of aircraft. These aircraft include fighter jets and attack helicopters. It takes many marines to maintain and use these aircraft. Some marines are pilots. Others are mechanics. Air traffic controllers have important jobs too.

Marine pilots fly helicopters and other aircraft. They must have a bachelor's degree. Some marines earn this degree from a nonmilitary college or university. Others attend the US Naval Academy. Only officers can become Marine pilots. Candidates go to flight school in Florida. They learn to fly different types of aircraft. Pilots must serve in the military for at least eight years.

Mechanics keep the aircraft in working order. There are many parts on an aircraft. Mechanics can specialize in maintaining and fixing certain parts. Some mechanics work on aircraft engines. Others learn how to maintain mechanical systems on an

Marine pilots learn how to use special navigation and communication equipment.

airplane. Marines can learn how to work on several types of aircraft. These marines may be put in management positions. They oversee a group of aircraft mechanics.

AIR TRAFFIC CONTROL

Many aircraft may be coming and going from a base at any given time. Some marines control the flow of aircraft traffic. They direct aircraft that are in the air and on the runway. Sometimes they have to control traffic in dangerous combat situations.

Air traffic controllers use different types of equipment to direct air traffic. Radar equipment sends out radio waves. The waves hit objects. Then they bounce back. Controllers can find out how far away an aircraft is. They can also see how fast it is moving. Radar helps them track

Marine air traffic controllers must be able to handle a lot of information and stay calm under pressure.

incoming and outgoing aircraft. Each plane shows up on a screen. The controller can see where each plane is in the sky. Controllers also use charts, maps, and other resources. They must understand meteorology. Meteorology is the science of weather forecasting. Air traffic controllers need to predict how the weather will affect the aircraft and routes.

Air traffic controllers talk to pilots through radio equipment. They must be able to speak directions clearly. They must be well organized. They have to manage

DOG HANDLERS

Some Marine police are dog handlers. They work with dogs. The dogs search vehicles, buildings, and open areas. They are trained to find illegal drugs and explosives. They are also trained to search for people. They can perform search-and-rescue missions. The dogs train for three years. Their handlers continue to train them afterward.

many things at once. Corporal Micah Rasmussen is an air traffic controller. He works at a base in San Diego. He explains the importance of his job. He says, "We're the [people] that coordinate everything that goes on in the sky."[5]

Many people are needed to help meet the Marine Corps' mission. They make the Marine Corps a fast and effective fighting force. Marines work in a variety of roles. Some have office jobs. Others prepare meals to feed marines. Marine military police enforce rules. There are many job opportunities within the USMC.

CHAPTER THREE

WHAT IS THE LIFE OF A MARINE LIKE?

A marine's daily routine often varies. Some marines are deployed. This means they are sent to serve in another country. Others live and work in the United States. There are about twenty Marine Corps bases, camps, and airfields in the United States. There are also many

Marines do training exercises to develop and strengthen the skills they will need.

Marine Corps bases overseas. Marines are stationed at these locations.

ON BASE

New marines who are stationed on a base live in barracks. Barracks are large buildings. They have common kitchens and living rooms. They also have smaller rooms. Each room has a small bathroom and bedroom area. Two people live in each of these rooms. Marines must keep their barracks and rooms neat and tidy.

Some marines are married or have families. On-base houses may be available for these marines. Higher-ranking marines may live off base in apartments.

Marine Corps bases have different facilities to meet marines' needs. Hospitals, churches, schools, and limited shopping are available. Marines can find restaurants as well. There are also many

Single recruits in all branches of the military live in barracks.

entertainment options. They may include movie theaters and bowling alleys.

DAILY ROUTINES

Many marines have daily routines. They stick to a certain schedule. The schedule depends on a marine's job.

For example, combat marines might wake up around 5:30 a.m. They have about thirty minutes to get ready.

They get dressed. They sweep and mop the barracks floor.

Marines join their **formations** after cleanup. They may organize into squads, platoons, or companies. Each marine is

> ### KEEP IT CLEAN!
> One night a week, marines focus on cleaning their barracks. This is called "Field Day." Officers inspect the barracks the next morning. It is important to keep these places clean. Unclean areas could have germs. Or they might have mold or mildew. These things can make marines sick. Michael Kagle is a sergeant in the Marine Corps. He explains, "More [marines] have fallen to disease and sickness in combat than actual bullets."
> *Quoted in Lance Cpl. Adam Johnston, "Continuing the Field Day Tradition," United States Marine Corps, March 22, 2006. www.marines.mil.*

part of a team. A team has three marines and one leader. A squad is made up of three teams. They report to the squad leader. A platoon has three squads. Three or more platoons make up a company. Leaders make sure everyone in their group is present.

Next, marines go through physical training (PT). PT includes different exercises. Marines may do push-ups and sit-ups. They may run in formation.

Marines get a break at around 9:00 a.m. They have time to shower and clean up. They dress in their uniforms.

Many marines later have classroom training. They may learn how to use weapons or special equipment.

Most marines have lunch around 11:30 a.m. They have a break until 1:00 p.m. Then marines fall back into formation. They may have weapons training. They may practice shooting weapons on a firing range. They also have combat training. This training ends around 4:00 to 6:00 p.m. Officers then share important news or information with the marines.

After dinner, the marines are dismissed. They can change out of their uniforms.

Marine Corps Formations

Type of Formation	Who Is in the Formation?	Total Number of Marines
Team	Three marines	Three
Squad	Three teams	Nine
Platoon	Three squads	Twenty-seven
Company	Three or more platoons	243
Battalion	Three or more companies	729
Regiment	Three battalions	2,187
Division	Three regiments	6,561

"Marine Corps Command Structure," US Department of Defense, n.d., www.defense.gov.

This chart shows the types of marine infantry formations.

They have free time. Some marines play video games or card games. Others go see a movie or hang out with friends. Marines must get a good night's sleep for the next day's PT.

Marines who live off base drive to the base in the morning. They arrive in time for morning formation. They drive home later when the workday is done.

ABOARD NAVAL SHIPS

Marines are often stationed aboard navy ships. This deployment usually lasts six months. Space on a ship is often limited. Marines live in close quarters. They sleep in bunk beds. Bunk beds are stacked three high.

Marines on naval ships may be sent to fight in another country. Or they may help with recovery efforts after a natural disaster. They need to be ready to respond at a moment's notice. They must stay in shape.

Marines practice drills when deployed on naval ships. They must be prepared to respond to an attack.

They lift weights and run on the deck. The marines do not have duties related to the ships. They are responsible for keeping their gear in order. Jhensenn J. Reyes is a Marine sergeant. He says, "Marines need to stay focused as we make our way to our destination. . . . At any moment we could be called on to do any type of mission."[6]

SERVING ABROAD

In wartime, marines may be deployed to dangerous areas. For example, some marines serve in Afghanistan. They face threats from roadside bombings and attacks. They talk to Afghans every day. Some Afghans are innocent **civilians**. Others may be enemy fighters.

Daily life during deployment may vary. Marines do not typically stick to normal routines. They must be prepared for any type of situation. Enemy attacks can happen at any time. Marines rely on their training so that they are ready for anything.

THE MARINE CORPS' SHOOTING TEAM

The USMC has a shooting team. People on this team are skilled shooters. They have good aim. The team participates in Marine Corps Marksmanship Competitions (MCMCs). MCMCs happen abroad and in the United States. Each competition lasts two weeks. The shooting team trains other marines. Team members teach shooting techniques. People fire at targets on shooting ranges. Major Mark O'Driscoll is an officer. He is in charge of the USMC shooting team. He says, "Competition breeds excellence."

Quoted in Calvin Shomaker, "Marine Corps Marksmanship Competitions Begin Dec. 9," The Globe, November 25, 2019. www.camplejeuneglobe.com.

CHAPTER FOUR

WHERE DO MARINES SERVE?

Marines may be deployed for many types of missions. They are sent to many parts of the world. They are deployed to areas that are in conflict. But they are also sent to non-dangerous areas. They provide security at US embassies. Embassies are in foreign countries. US government employees work in them.

There are US embassies in almost every country. Some marines protect these buildings.

FIGHTING TERRORISM

On September 11, 2001, **terrorists** hijacked four airplanes. They flew two planes into the World Trade Center towers.

These buildings were in New York City. A third plane hit the Pentagon. This building is near Washington, DC. It houses the US Department of Defense. The fourth plane crashed into a field in Shanksville, Pennsylvania. The attacks killed nearly 3,000 people. The terrorists were part of a group called al-Qaeda. This group was based in Afghanistan. It was led by Osama bin Laden. In response to the attacks, the US military invaded Afghanistan. Its goal was to remove bin Laden from power. US Navy special forces finally killed bin Laden in 2011.

Thousands of marines are deployed throughout the Middle East.

The War in Afghanistan continued into 2020. From 2010 to 2011, about 20,000 marines were deployed to Afghanistan.

That was at the height of the war. More than 114,000 total marines have been deployed to Afghanistan since 2001. Some were redeployed there as many as five times. Today, many marines remain in the region. They advise local governments. They also

ANNUAL TESTS

Marines have to pass two fitness tests each year. One is the Physical Fitness Test. It involves many exercises, including push-ups and pull-ups. Marines must also do a 3-mile (5-km) run. The run is timed. Marines have to meet certain time requirements. The requirements depend on their age and gender. The second test is called a Combat Fitness Test. It involves a run and an obstacle course. Marines lift ammo cans too.

provide security to keep neighborhoods safe. Terrorist groups are still active in Afghanistan. Marines help defend people from terrorist attacks.

Marines also support the residents of Afghanistan. Sergeant Lauren Nowak leads a team of marines. She serves in Afghanistan. Her team built a school to teach local children. She explains the importance of this school. She says, "[Students] can . . . help their family through education. They can help themselves. It gives them a lot more options in their future."[7]

Marines have also fought terrorism in the Philippines. Terrorists took over the city of Marawi in 2017. Philippine soldiers fought to take back the city. The marines advised and assisted them. The battle took five months. The soldiers were able to take back the city. Today, the marines continue to help train the Philippines' military.

DISASTER RELIEF AND BORDER SECURITY

Another important Marine Corps mission is disaster relief. In 2019, 200 marines were sent to Central and South America. Hurricanes had hit the area. The marines' mission was to help people recover after the storms.

Marines also train with troops in other countries. They can learn from other countries' military forces. They can also teach these troops important skills. In 2018,

TRAINING IN SCOTLAND

Marines must learn important survival skills. Some marines train in Scotland. Scotland has a cold climate. Rain is common there. Marines learn how to hunt and build fires in these conditions. They also learn how to navigate. Graham Perry is a major in the Marine Corps. He says, "We typically find ourselves in [extreme] conditions and environments. . . . Survival training will help the marines."

Quoted in Corporal Dallas Johnson, "Marines Learn Survival Skills in Scottish Highlands," US Department of Defense, *April 30, 2018. www.defense.gov.*

In 2018, marines deployed in Peru showed local troops how to use special weapons and equipment.

marines were sent to Peru. They trained with local troops. They learned how to work together to help with disaster relief.

Other marines help with border security. They are stationed along the US–Mexico border. They patrol the border. Their goal is to keep people from entering the United States illegally.

CONTINUED TRAINING

Marines continue training throughout their career. Sometimes they go to other parts of the world to train. They may train in Alaska. This helps them prepare for cold weather warfare. Marines also train in

Some marines take a cold weather survival course at Fort McCoy in Wisconsin. This prepares them for deployment in cold areas.

northern Norway. They learn how to survive in arctic areas. These places have harsh weather and landscapes. Marines learn how to adapt to many environments.

DEPLOYED MARINES AND THEIR FAMILIES

Marines who are deployed often do not have regular bedrooms. Instead, they usually sleep in tents or on the ground. They may not have running water. They may have to use bottled water. They wash their laundry with bottled water and soap. They cook their food over an open fire. Meals are fairly simple. Some marines use ingredients that they buy from local people.

Marines keep busy while deployed. They take part in military missions. Sometimes they have downtime. They may watch movies on laptops or call their families. Some play cards or darts. Others read or play sports. They also work out to keep in shape.

A deployment can last several months. Marines do not come home during this time. They may not see their families for a while. This can be difficult. Marines can video chat with their families and talk on the phone.

There are many resources to help Marine families. Support groups provide emotional support. Some organizations host events for military families. Others offer opportunities for children, such as art camps. These activities help keep families busy during long deployments. Families can talk to others who are going through similar experiences. Many families make videos to send to a parent who is deployed.

COMING HOME

It is exciting when marines return from deployment. Families gather to welcome the marines home. They make signs with words of welcome. They might dress in red, white, and blue. They might also wave small flags. Sometimes, families do not know their marine is coming home. Lance Corporal Justin Burgos surprised his family. He walked out onto the field at a New York Giants football game. He had been deployed for more than a year.

Being a marine is difficult and requires a lot of training, but the work can be rewarding.

Some put together care packages to mail to their marine.

Serving in the Marines is a commitment. It requires hard work and dedication. Marines are loyal and fierce fighters. They face many challenges. But they know their service is important. They work every day to protect and defend the United States.

GLOSSARY

ammunition

a supply of bullets and explosives

civilians

people who are not in the military

formations

groups into which marines are organized

infantry

troops who fight on foot

recruits

people who have recently entered the Marines but are not yet fully trained

stealthy

quiet and secretive

terrorists

people who target and attack certain people or groups in order to spread fear

SOURCE NOTES

CHAPTER ONE: HOW DOES A PERSON BECOME A MARINE?

1. Quoted in Graham Flanagan, "We Went Inside the US Marine Corps' 13-Week Boot Camp," *Business Insider*, October 30, 2019. www.businessinsider.com.

2. Quoted in Lynsey Addario, "Women Becoming Marines: 'I' Will No Longer Be in Your Vocabulary," *New York Times*, March 24, 2019. www.nytimes.com.

3. Quoted in Flanagan, "We Went Inside the US Marine Corps' 13-Week Boot Camp."

CHAPTER TWO: WHAT JOBS ARE AVAILABLE?

4. Peter James Kiernan, "5 Things I Learned from the Marine Corps' Scout Sniper School," *Task and Purpose*, June 13, 2016. www.taskandpurpose.com.

5. Quoted in "A Day in the Life of a Marine Air Traffic Controller," *Marines*, December 6, 2012. www.marines.mil.

CHAPTER THREE: WHAT IS THE LIFE OF A MARINE LIKE?

6. Quoted in Lance Cpl. Santiago G. Colon Jr., "Living on Ship: 26th MEU Marines Balance Duty and Downtime," *Marines*, September 15, 2010. www.26thmeu.marines.mil.

CHAPTER FOUR: WHERE DO MARINES SERVE?

7. Quoted in "Interview with Sgt. Lauren Nowak," *Eielson Air Force Base*, November 29, 2011. www.eielson.af.mil.

FOR FURTHER RESEARCH

BOOKS

Roberta Baxter, *Work in the Military*. San Diego, CA: ReferencePoint Press, 2020.

Kristin J. Russo, *Surprising Facts about Being a Marine*. North Mankato, MN: Capstone Press, 2018.

Jim Whiting, *Marine Force Recon*. Mankato, MN: Creative Education, 2018.

INTERNET SOURCES

Philip Athey, "Thousands of Marines with 26th MEU End Exercise, Head to Mediterranean Sea Amid Rising Middle East Tensions," *Marine Corps Times*, January 6, 2020. www.marinecorpstimes.com.

Lance Corporal Santiago G. Colon Jr, "Living on Ship: 26th MEU Marines Balance Duty and Downtime," *United States Marine Corps*, September 15, 2010. www.26thmeu.marines.mil.

Shawn Snow, "The 13th Marine Expeditionary Unit Is Deploying to Central and South America," *Marine Corps Times*, September 18, 2019. www.marinecorpstimes.com.

WEBSITES

National Museum of the Marine Corps
www.usmcmuseum.com

The National Museum of the Marine Corps is in Triangle, Virginia. Visitors can learn about the history of the Marine Corps. The museum's website offers information on exhibits and events.

Ready Marine Corps Kids
www.ready.marines.mil/Ready-Marine-Corps-Kids

This website is for the children of deployed marines. It offers activities, games, and challenges. These activities can help children deal with the stress of a parent's deployment.

The US Marine Corps
www.marines.mil

This is the official website of the US Marine Corps. It has information about how to join the Marines. It also answers questions about what it is like to be part of the Marines.

INDEX

Afghanistan, 56, 60–63
air traffic controllers, 34, 39, 42–45
aircraft, 33, 39–44
Armed Services Vocational Aptitude Battery (ASVAB) test, 20–21

barracks, 47–48, 50
bases, 16, 32, 42, 45, 46–49, 53
bin Laden, Osama, 60
boot camp, 4–5, 21–31, 32

combat, 5, 10, 18, 24, 29, 35, 42, 50, 52, 62
Crucible, the, 29–30

deployment, 46, 54, 56, 58, 61–62, 69–71
disaster relief, 8–9, 33, 54, 64–65, 67

embassies, US, 58
enlistment, 13–15, 18–19

infantry, 33, 35, 37
Initial Strength Test (IST), 24–26

leadership, 13, 18, 51

Marine Corps Recruit Depots, 22–23
Marine Corps Reserve, 4, 12

National Reserve Officer Training Corps, 16
naval ships, 8–9, 32–33, 54–55

Officer Candidate School (OCS), 16

Parris Island, South Carolina, 5, 22–23
physical training (PT), 51, 53

recruits, 4–5, 18–31
requirements, 14–15, 20–21, 25–26, 36, 62, 73
riflemen, 34

San Diego, California, 5, 23, 45
scout snipers, 35–37
survival skills, 24, 27, 65, 68

terrorism, 59–60, 63–64

US Naval Academy (USNA), 16–18, 40

IMAGE CREDITS

Cover: Cpl. Jesula Jeanlouis/U.S. Marine Corps/DVIDS
5: Lance Cpl. Nathaniel Hamilton/U.S. Marine Corps/DVIDS
7: Lance Cpl. Alexis C. Schneider/U.S. Marine Corps/DVIDS
8: Mass Communication Specialist 3rd Class Dana D. Legg/U.S. Navy/DVIDS
11: Cpl. Harrison Rakhshani/U.S. Marine Corps/DVIDS
13: Warrant Officer Bobby Yarbrough/U.S. Marine Corps/DVIDS
14: Warrant Officer Kowshon Ye/U.S. Marine Corps/DVIDS
17: Lance Cpl. Piper Ballantine/U.S. Marine Corps/DVIDS
19: © Keri Delaney/Shutterstock Images
20: Sgt. Erica Kirsop/DVIDS
27: Lance Cpl. Samuel Fletcher/U.S. Marine Corps/DVIDS
28: Lance Cpl. Samuel Fletcher/U.S. Marine Corps/DVIDS
31: Chief Warrant Officer Bobby Yarbrough/U.S. Marine Corps/DVIDS
33: Lance Cpl. Terry Wong/U.S. Marine Corps/DVIDS
37: Lance Cpl. Samantha Sanchez/U.S. Marine Corps/DVIDS
38: Cpl. Michelle Reif/U.S. Marine Corps/DVIDS
41: Cpl. Seth Rosenberg/U.S. Marine Corps/DVIDS
43: Cpl. Noah Rudash/U.S. Marine Corps/DVIDS
47: Warrant Officer Kowshon Ye/U.S. Marine Corps/DVIDS
49: Petty Officer 2nd Class Charles Oki/U.S. Navy/DVIDS
53: © Red Line Editorial
55: Lance Cpl. Joshua Sechser/U.S. Marine Corps/DVIDS
59: Sgt. Kyle Talbot/U.S. Marine Corps/DVIDS
61: Cpl. Rhita Daniel/U.S. Marine Corps/DVIDS
66: Staff Sgt. Frans E. Labranche/U.S. Marine Corps/DVIDS
68: Scott Sturkul/U.S. Army/DVIDS
72: Sgt. Audrey Rampton/U.S. Marine Corps/DVIDS

ABOUT THE AUTHOR

Susan E. Hamen has written many books for young readers. Some of the topics she has covered include the Wright brothers, the Civil War, and ancient Rome. She lives in Minnesota with her husband, daughter, and son. Together with her family, she likes to travel, play music, and experience new things every chance she can get.